The Cartographer's Vacation

Owl Creek Press
2693 West Camano Drive
Camano Island, WA 98292

(360) 387-6101

FIRST EDITION

The Cartographer's Vacation

POEMS

Andrea Cohen

Owl Creek Press • 1999

I am grateful to the editors of those magazines in which these poems first appeared: *Crazyhorse*, "Gogol's Madman"; *Cumberland Poetry Review*, "Roll Call, Between the Coups"; *The Denver Quarterly*, "Segue"; *High Plains Literary Review*, " The Cartographer's Vacation"; *Kentucky Poetry Review*, "Curse of the Uninvited"; *Orion*, "Evidence"; *Ploughshares*, "Story of the Tattoo" and "Ode to the Noodle"; *Poetry Northwest*, "Mild Instructions for Travel."

"The Possible World Times Seven" and "Reinventing the Wheel" were included in the compact disc, *One Side of the River* (Say That! Productions).

My gratitude to those who have been my compass in this and wider circles: Amy Anderson, Gigi Munafo, Francesca Bewer, Toni Lee Pomeroy, Greg Pomeroy, Sarah Keniston, Naomi Wallace, Karen Bernstein, Jean Wilcox, Rita Halbright, Bob Steinberg, Lise Motherwell, Emily Hiestand, and my brothers, Bruce and Steve.

For my parents
Dorothy and Sheldon Cohen

Contents

People possess four things
that are no good at sea:
anchor, rudder, oars,
and the fear of going down.

ANTONIO MACHADO

Details You Must Remember

The snowflake begins with a premise.

The white road is the waiting
room for the forest.

Stars travel days and nights.

You must imagine yourself
as a place you will visit

one unforgettable afternoon,

running in from a spring rain
for small cakes and sweet wine.

The swallows explain this

in textbook skies, as does the wheel
without end, which, traveling, keeps

one eye turned backward: the hibiscus

garden you lie down in
is not the garden you are leaving.

Every hand cupping water to your lips is direction.

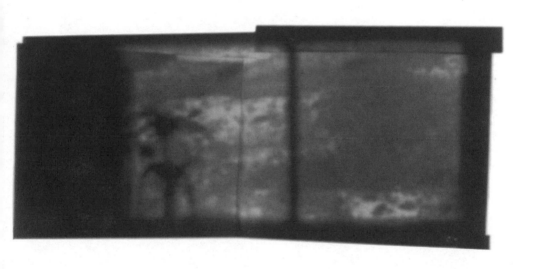

I

The Physical World Times Seven

Father's bare-chested on the autumn rooftop,
tinkering with brittle equations
from a high school geometry book
and an experiment involving gutters.
My mother stands at the foot
of the ladder, at the pinnacle
of her beauty (which we can't know then).
Her voice ascends like nectar
in the mouths of hummingbirds,
and as swiftly as rainwater
he descends and vanishes inside her arms.

Meanwhile, I'm constructing
a tilted fort between three elms,
inside the fragrant rain
of chestnuts, through which
the fat squirrels hurry.
I am eight years old.
It's the day before the lovesick neighbor
steals the velvet blooms
of red camellias
climbing our stone wall—
though he is never accused
of anything and my father explains
that *these things happen*.

It is the autumn of my figuring
everyone's age in dog years,
rendering all I love
possible beyond our merest bodies.

Memory

Quietly the snow begins
falling filling
the windowsill
like a glass
of milk our mother
is pouring

How long ago
she began
before the gardener
reclaimed his forsythia
before the ants carried off
the chairs and lamps

We expected her task
to take forever
oh if she could see us now
surely she'd come back
to scold us
as the auctioneer
carts off the whole house

save the one window
we kneel before
fingers plugging our ears
fending off the shackling
laughter of the wealthy snowmaker
and his machinery
for which we pawned
our tongues
for which we'd pay anything

Heart Grown Fonder

Mother relished drama,
whispering rumors to the ant farm
to get everybody going.
On Saturday afternoons we dressed
in sample perfumes and carried
frayed satin stools
to the living room window.
Every performance is different,
she'd whisper.
I know this now.
The wind chose different victims
daily, migrant clouds darkened
random girls and rooftops,
and Lily the cat made herself
invisible in the root cellar
or behind the stone wall.
I never knew what
new absences might turn up:
the mailbox my father
took with him, the white rose thicket
whose every petal of a sudden
searched a swooning couch.
We watched one autumn twilight
as six broad-backed hoodlums
hoisted the barn
from its foundation.
Wild phlox and shepherd's purse,
purple loosestrife and woodbine
grew in its place.
Even now mother waits
for the curtain's ascent.
Liefert, she says to me,

murderer, cream puff,
thief, you shyest of toads.
I'm her dear Uncle Lonely,
Sir Fetch-me-my-glasses.
She takes my face
in her wilting hands,
kisses Tuesday, spits Thursday,
love's rutted parade
come to roost on my shoulders.

Oh, I Do Not Believe in Original Sin

I am seven years old
and wander into a crowd
of temptation where
the sultry fortune teller,
sad-faced rabbit girl
and two-legged lamb all beckon.
Gold-toothed carnies
promise unlimited prizes,
and even the poodle-shaped pumpkin
whispers stark invitations.

It is getting dark out.
I am riding the Ferris wheel.
Night clouds lumber past,
a parade of colossal,
bumbling, helium animals.
The lonely barks
of the seal-boy float up—
and I bark back.
I explain to myself that I am
not lost, but circling.
I am very convincing.

By the time my father finds me
so much has yet to happen.
The leaves are grinding
their teeth in the wind.
I am lost and something grave
is gone wrong with the world
as he takes my hand
and the echoing loudspeaker
announces my checked shirt, blue tennis shoes,
and dire predicament.

Headless Heed

During our standard-issue Saturday lunch of Vienna
sausages and root beer, my older sister Agnes whispered to
me that the school janitor had found a woman's head in the
girls' room commode. I was six years old. That evening over
pot roast, she confided that the school principal had later
found the janitor's head in the little boys' room. I did not
speak to my sister after this confession. However, I did
begin to treat the swinging doors of public rest rooms with
the perverse and ginger caution Pandora must have exer-
cised when tempting the lid of her infamous bin. I knew
that a head could appear in the lavatory of a jetliner speed-
ing above Minneapolis or in the pink faux-marbled facilities
of Miss Munafo's Modern Dance Studio. It might surface in
the musty, oak stall of the public library or in the sparkling
bath room at the pediatrician's office. But doubtlessly, it
would turn up somewhere, rooted, like all evil, on the tip of
Agnes' cruel tongue.

The gruesome head tugged at my imagination like a mighty
terrier at the end of its brief leash. Knowing the worst that
lay in store, I was strangely prepared for everything.
Throughout school I excelled with minimal study. Answers
leapt from my pencil. I had no choice but to prosper as a
world-renowned physicist.

My sister, on the other hand, was a profound failure. She
had peaked with an A+ in recorder in the sixth grade, and
declined from there. At the height of Agnes' failure, she
became a poet—a horror that was little discussed in our
family. But Agnes was unashamed of her descent into verse.
On birthdays she showered us with sonnets. Funerals
prompted sestinas, and each Halloween, Agnes wired a
prose poem, which confounded us further.

This continued for thirty-seven years, during which time I continued not to speak to Agnes. Despite my silence, she invited me to a reading of her villanelles in a coffeehouse in Amsterdam, where I happened to be delivering a paper on the physics of underwater propulsion.

My paper was well-received. In the question and answer session, I provided one answer for each question. Afterwards, the chubby, young daughter of a shipping tycoon presented me with a dozen roses.

I hurried through a thunderstorm with the flowers to Agnes' reading in a grim and grimy part of the city. Inspecting the roses more closely, I found that each bloom, now wilting, had been wired to a stem. And it occurred to me that to come upon a head in a commode, one would rightly encounter a body first, or at least a trail of blood. No matter how removed the head might be from the rest of itself, hints to the prior connection would be impossible to mask.

By the time I reached the coffeehouse, my sister was just finishing up. For the first time, I realized how small and turtle-like she was, hidden behind the lectern, only her bobbing head visible. Although the acoustics were terrible, I could make out that she was a waitress reciting the day's specials, each more succulent than the last. For the pièce de résistance, Agnes' heart leapt through her lips onto a golden platter.

It was a tremendous heart. Outside, snow had begun to fall and the canals were heading home. On the back of a cocktail napkin I wrote Agnes an anonymous fan letter in Dutch, a language I had never spoken before, praising the tender genius of a sister I never had, the broken, beaming bouquet that was our birthright.

Family Portrait

Father with gorgeous Regret
stripteasing on his shoulder,
my paroled kid brother
filching the tenement stairs,
and dear old Mom
knitting tea cozies
for the neighborhood rooftops.
That's me with a musty book
in the dim corner,
rereading the fairy tale
in which we all appear
briefly as ourselves
before boarding the ocean liner
with forged passports, mismatched shoes,
and happy hearts
tattooed to our eyelids,
which naively flutter
as we sail the unlikely
red carpet
of sunrise
upon unsaddled sunrise.

Indian Summer

The laughing boy in striped shorts trots
his white puppy down the morning avenue.
Already, the lost dog, the masked one they called
Bandit, Bandit, Bandit from the front porch
and dark car windows through desolate patrols
in foreign neighborhoods is forgotten.

He tumbles into a soufflé of raked leaves
and plucks one fiery orange, five-fingered maple leaf
which, on the flagpole of his small arm swaggers
the warm breeze as he pushes it toward the postman
as one might present a theater ticket
to the usher of a renowned production.

He waves the leaf to the creature at his heels who
bellies up, surrendered like some thrice-smiling, dutiful
Cerberus, mesmerized, as from summer's verdure
a staggering splendor rises.
Amidst such miracle alchemy, even Cerberus wags his dragon's tail,
certain that raising the dead's nearly plausible.

The sun climbs a ladder of blinded stars, the boy
forgets his stalk, lets the open gate swing on its hinges,
blurring in the distance as they round the corner,
the white dog picking up speed, passing the boy,
a sentry racing ahead, then back, then into the woods
to bury fresh bones and reclaim fossils.

Around them, the hourglasses of oaks and maples
keep their watch. Each autumn, the county fair promises
both prize radish displays and a glimpse at the snake-girl
hissing fork-tongued behind her reptilian veil.

And each spring the monarch butterflies migrating to milkweed
sweep the sky, like a velvet curtain closed, then opened.

When the white dog dies, the boy will call
the animal his childhood, and the death of it.
Shadows will lengthen, the world will turn
to fire and ice over and over, and yet—
how but love what inhabits us,
though by design it falls from our most ardent grasp?

Speculation maps the meadow of ghostly asphodels.
Today, sunlight defies the season and family hounds bark
their allotted length of picket fence
and unharvested pumpkin fields, and above them the great elms
trail the leafy, rainbow scarves amidst whose finery
one forgets to look back, and no love's irretrievable.

Segue

Some tribes red ochre a child,
the whole lifeless body berry red
except for the face, unlined, genitals
unversed in anything
other than daily functions.
At those two places the red dye
peels away, like death embarrassed,
like death, on second thought, confronting
an error that grows more hideous
as more eyes disrobe it.
The legend says the two circles
of bareness allow the child, once unasked,
to choose gender in the next life.
I suppose that is how
we living bear ourselves.
In New Guinea a boy has his ankles
strapped to vines as tall as plantains,
climbs a platform two petals taller
than the vine, and lunges. Sometimes
the vines don't fray, and he smiles, suspended
the breadth of two petals above ground.
Sometimes the vines can't hold
him to this world: that fatal Achilles
flaw that has nothing
to do with a child. A rite
of passage. But there is no word
for reason in such cases,
in any language.
A boy is born with a hole
in his heart, like a star
blazed through his chest.
Oh, but a star

is no consolation for the dead.
It's not as if blossoming
as a boy next time, or a girl,
will be better.
Why not ladle hibiscus
around the boy's ankles,
rooting him to this earth,
his whole body unstained,
and let that be his passage.

Gogol's Madman

The sky's not blue nor falling, nor
treasured hands from an open window beckoning.
We never dressed properly, rose hips
never crowned our unroyal heads nor
any feature of the stony day, the sky
not falling, the great sea booked for passage,
the trail of islands strewn like bread crumbs
in the sea's dissolve, the way home
never was, the way the precipice always
crumbled, a drumming
dirge into that rising swell—

Say otherwise. That we were there,
and happy, that possessing the correct
scripts, we got all the best parts.
Our hair swooned into place,
our daydreams were copyrighted,
our limbs made communion with the moon and
sometimes we laughed a little. Didn't
the dinner bell ring all day and the cassoulet
tend itself? Was not the mass exodus
of trumpet players a ruse and returned,
don't they play still? Didn't the sky
invite us up into it, our ascent
the envy of Spain? Didn't we laugh
at the right moments, wasn't the hand
waving this way and weren't
we well-frocked and hasn't she been there
all along, my adoring mother, still luxuriant,
gathering me up with her milky arms
into the stars, where I reign,
Ferdinand VIII, supreme, the only and best

King of Spain, under whose measured rule
all dine on emeralds and blue wine
and at long last the correspondence
of dogs rings true, unencumbered
with denial or treachery?

Odyssey of Many Returns

It was an impromptu excursion,
without worry beads
or royal ornithologist
to paw the rare aviaries
of dreamy locales.
(The itinerary hadn't dried,
and so remained home, a ward of the houseflies.)
Like anything, the trip began
as a series of tests:
deep-sea diving for remote
control devices, slaying sun
glass lenses for rented dictators,
confiscating goat milk from Guernseys
and sawing a stunning apostle
into cucumber sandwiches.
I was ice-fishing for golden-tipped arrows
in Lone Tree, Iowa,
when a corps of defecting,
pink crab apple blossoms
executed Swan Lake's finale
across the melting pond.
I could barely hear my own applause
for the bravos of a visiting parrot,
and was just about to seize its gibberish beak
for my trophy room when up the horsehair line
my blue-lipped father climbs,
shivering and scolding.
How long it had been!
He whistled the Marsellaise with the parrot.
There was always so much
I didn't know about him.
But straight away he eyed

my brimming pockets
and dutifully I took his hand,
and one by one we made the rounds
returning every last doorstop
to its slack-jawed, knee-slapping proprietor.

Icarus Obsessed with the Possible

The terrible thing about this world is that everybody has his reasons.
JEAN RENOIR

It isn't vaguely sexual
the cicadas grating their huge hind legs
they've hidden in the elms and red maples
looking like leaves laughing
at an inside joke
which is us

the long cord of the venetian blind moves
because of wind because the wind
moves the windmill supplying power
enough to grind grain to feed us
because in this fashion we live long
enough until tomorrow

I am walking down a street of bricks
how can I make the street
look like 1928
remove that boy's clothes those cars
replace that stained glass
why should I desire an appearance
because I am a film director
with a small budget and poor imagination
because in the scheme of things
moderation is little tolerated

obsession is both the brass ring
and my desire
if I believe I'm justified
in stealing the ring
or slinging that boy into the river
will charges be pressed

is the mere existence of the object
my implied doom

I've got shades of fingernail polish to compliment
whatever falls into my hands

do we always want dead what's drawn
to light consider Icarus
as we've been trained to expect him
why Icarus lives just beyond the 5 & dime
Icarus delivers evening papers
and when Icarus falls it's our failing
since time immemorial but also a moral
valid toward reprieve
and Icarus that blonde boy is not even the enemy
God help us
consider the housefly
its title implies domain
across the dark and sleepless lawns the flies play
rods of purple electrical current
a symphony and the sound is nothing
human unless the sound
of a bitter mother calling
her bitter son
already out of earshot

Curse of the Uninvited

The downstairs guests spin records,
laughing inordinately. I latch the door, lay
the mattress on the bare floor and turn my head
to the window, eye-level. Outside:
the gravel alley, oaks, stone garage, and sorry needle
of a moon. A frantic housefly cases this side
of the screen meant to exclude him.
Insatiable legs of ballerinas.

How best to direct you
to the front door, its occasional openings?
Aligning yourself with the screen
only multiplies angles beyond your reach,
what now seems fly-heaven outside.
One hope is happening on
the gap meant for rainwater's escape.
Weather's also an intruder here.

If you stay still you might see
your predicament entirely. *Shhh*.
Once, in the land of twelve golden plates,
the thirteenth fairy, uninvited,
arrived, cursing everyone to stone, yes
even the fly on the wall. The tradition
holds a savior came, a prince
after one hundred years, but there is room
for speculation. Perhaps briar roses
still contain the kingdom of stone.

At least you and I can touch
the jagged screen, the cool door returned
to its frame. We should count
our blessings.

Outside, a bone of a girl
I met once, moon-faced, roams the streets
in sackcloth, as if waiting
for rescue from Dachau in America,
her head shaved of her own volition.

By the end of the story people live
as happily as possible.

Winter of My Youth, with No Address

It was a bad winter
and I took the job
clipping the eyes from needles.
I got used to finger pricks
and the boss left me alone,
stopping by in the late afternoon
to collect the fruits
of my dismemberment.
He'd ask about mail
and I didn't have the heart
to tell him the makeshift office
lacked an address,
which was why I'd arrived
weeks late for the initial interview.
I could sense his beautiful wife
had moved in with a matador
and taken the salt shaker.
At closing I tiptoed past his closed door,
at whose foot a slim crop
of light lay,
like the dried shavings of a lemon.
His tender scolding
of the eyes and their adjacent
bodies followed me
all that bleak season:
This won't do at all,
things have got to change,
and that scent so delirious, epoxy.

Story of the Tattoo

As I recall, it was night in another country. The neon
city puckered with fuschia, aqua, amber, and cobalt,
and bare-chested men were shattering windows in
prayers for some delicious, slight breeze, and small,
antiquated fans spun everywhere, making silence an
exotic and far-flung resort. As I walked, a heaven of
rain fell for two blocks and steam rose from the side-
walks and scattered pools of water gleamed already like
shadows.

Things can fall away from themselves that quickly.

The way I slid the back of my hand across my cheek.
It was someone else's wrist. She was so sweet, I think
to myself. What did she know, sweet girl?

I went in because buckets of ice were piled high behind
that one storefront window still intact. I was a young
girl. There may have been some stars in the offing, or
a moon even. He was not a deceitful man. He said it
would involve some pain. The ice was as sweet as
sleep. I knew that stars must be tiny icebergs. I closed
my eyes. *It is so simple,* he said. *Little scars that swim
together and carry you along, make you beautiful. This, you
never lose.*

I saw myself swimming beneath a lotus tree that
matched my every stroke. Then I was the lotus. I do
not remember where I slept that night.

I forget so much. I am afraid to sleep. Now, I am alone
in a pale room with no ceiling save ochre leaves, a

cranberry sky, and an occasional tourist in a hot-air balloon. Even naked, I wear the indigo bracelet on my left wrist where surely once I must have worn a time-piece.

What approaches my heart is always in transition.

I bathe in sea water and caress every place within reach. You cannot imagine the pain. If I dream, my tongue's awash with blossoms. I could write a cook-book about the sky. I imagine a traveler who might drop down and not ask directions, one to whom I'd tell my every secret. And still there would be more.

Today, I am located in a charcoal-gray region of the world. Overhead, birds of every color fly off with grace and direction. How did she get away, ebullient blue-veined flower, she who bleached my bones but would not steal desire?

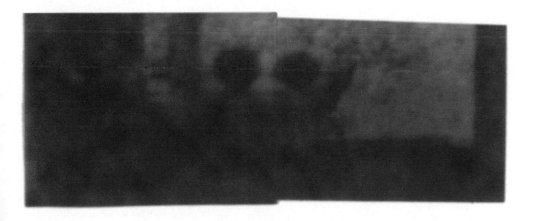

II

Reinventing the Wheel

We were saints
of the swoon and unbutton
in that fabled little seaside village.
After supping on peppered dogfish
and ouzo, we walked the dark
dirt alley home
where the white trumpet
blossoms of a showy catalpa
studded our path
like fragrant, fallen moons.

Beyond the aureate spleen
of memory, some lithe echo
persists, like the angel
always anointed
one town over.
Beyond memory, one is always
lost and re-beginning,
ever bedding down
and waking in darkness.
Always loping after the ancient
Chinese wisdom of the peasants:
When entering the dark,
first throw a stone
to figure, roughly,
where the black road lies.

Absent Host

While generally discarding such notices,
we accepted the invitation
to Eden, packed lightly,
and arrived on Tuesday.
Prepared by centuries of hearsay,
we were surprised to find
a tilled town in Iowa
with florists and pharmacies and green grocers
well stocked but mysteriously unchaperoned,
and each awning expertly mislabeled.
The shoe repair shop sold oriental rugs.
The carpet emporium offered perfect persimmons,
and the library stacks brimmed with bromeliads.
We had the whole joint to ourselves,
which meant no one spoke our language,
so we unlearned what we knew,
making love in corridors marked poison,
dining on red and redder wine
in a seaside cafe
marked surgical operating theater.
We strolled along a lane
dubbed History, where nothing
happened save the sense of lilacs.
We studied a collection of clouds
in a squat warehouse marked cemetery.
Everything was ours
indirectly, in the land of plenty—
shoes and shoeshines,
flotsam and jetsam,
but truth to tell,
we missed Cain and Abel.
We missed shadows and slot machines.

We longed for the narrow
escape from train wrecks,
passage on the leaky skiff
inside the perfect storm.
What could we do but cast
ourselves out, ill-frocked and penniless,
leaving the way
one exits a dream
of fruits singularly
succulent and indigestible,
greeted by that wondrous harangue,
the backlog of dust and tragedy.

In the City Where I Make Your Acquaintance

You kneel in the backyard
at the end of the world,
the turned earth clueless
as to what you pocket
in its crumbling lips.
In candlelight and the cold light
rain of late April,
you root the uprooted
strawberry plants escaped
from a victory garden.
Displaying a plant in each hand
and the cord between them,
you explain the nature
of proliferation:
how, shooting runners,
they tunnel
beneath the wire fence.
By reaching,
they multiply:
it is how we
strangers recognize each other,
in mutual perusal
of some shattered sky.

We may never know what griefs
exactly our hands mull over,
in commerce, in exchange
for a race of blossoms.
Scanning the scene you may believe
I am starting to define us,
but a gather of stones, scars, spoons,
is only the trail

from where we began.
Leaving, or returning,
we are not
the figures who entered
the plot of tilled land,
hardly yet a garden.
Already, the moon ladles out the night—
there is no apology for history,
only the flickering of these brief voices.

I Love You in Green Peas

Je ne saurais plus ou me mettre.
APOLLINAIRE

My eyes take a little time
adjusting to attic light and just
now you must have found
the *I love you* scripted in green peas
beside the stove.
There was no other way to put it.
I flooded the house
with wildflowers, anointed every
tabletop and dresser with bleeding hearts
and redbud and arranged jewelweed
on the bathroom linoleum.

From up here I can see
the garden plot you mulled over:
the neat rows of squash, their leaves
like ears of lop-eared rabbits,
the carrots airing like new whiskers.
And if legumes are akin
to small, defenseless animals,
should you leave them now,
leave this house
with instructions for me
to wake each day
and transplant the vegetables
as they overcrowd?
How will I know
when their density
is too much? Which seedlings
shall I move, and is the procedure

delicate, dangerous, as, say,
transferring goldfish? Where do I
place the new garden? Near the stone
garage? The oak woodpile?
It is so simple
that I love,
that I have many questions
and must detain you.

Citta d'Incanto

It is bigamy to love and to dream.
                    ODYSSEAS ELYTIS

Greetings, he sends, from a panorama
of the city of dreams, where behind abundant
bougainvillea, two men lean on the carved railing
above the many-bridged greenness of the Arno,
where red-clay tiled roofs give way
to the Pratomagno mountains shouldering the city.
Aside from the shapes
of mullioned windows and bridges,
and the suspicion that such vicissitudes
of lavender in this bougainvillea,
must be cultivated, one might never suspect
human goings on down there.
In the city of dreams,
the men must be shaking their watery heads, breathless
on account of a relic.
They might muse of interrupted love
down there, of delicate, white cakes,
and the church chiming once for each
soul having dragged the city with an outcast eye.

Further south, in the city of smoke
sent aloft, men make progress with the angels.
A new technology informs the actual intent,
in color, of Michelangelo's great ceiling, ergo,
reparation is possible, and modern genius
can preserve its predecessor: We are more
than short-term heroes in an historic setting.

Tourists flock to the chapel and cafes.
Light looks more important through the green

and magenta glass of Mary's cheek.
A girl in her pink dress turns, as instructed,
her back to the source of dreams and tosses
a coin over her shoulder into the fountain,
where the dreams of lusty travelers pool and sink.
With obsolescence built into the body,
threatening daily migration into the spirit,
one might pray for many deliveries.

Possible too, that as we murmur,
indecipherable messages travel here from other galaxies,
and that similar missives landed long ago
in murky swamps where marble-eyed lizards slithered.
We might also believe in local legend,
that once, where the Madonna delle Grazie
now stands, a massive team of oxen knelt in prayer.

Today, the sky's big-headed and blue.
In places, a vein of cloud forms a fissure
like the cracking in Adam's studied, outstretched
self. Was decay built into the portrayal
of longing? Fingernail, bone, lips poised
on the brink of prayer or embrace:
As the golden flesh-tones of the hand
flake into our upward gaze, do we gather
a language that prolongs beauty or defies gravity?

If the inhabitants of the city of God
abandoned their home,
this would be heresy.
If I left
your heart alone
it would be treason.

Variable Mountain Passes

I.

       Crossing the sweltering midlands, dozens of years
from an ocean, you devised
       a myriad of fortune-telling games:
if the next song on the radio,
       if that blue car in the rear-view,
if those cornfields, if that cloudburst,
       if that green pond, unrippled
as, if our clothes, if we, if, I—
       your dead reckoning
rendered every road possible.
       Heading east, we got lost,
drove too far north and had to travel
       a dozen narrow mountain tunnels
whose hypnotic white tiles
       almost had us believing
we could swerve and land
       in another world
without pain or loss.

II.

       We are worth our weight
in clichés, in flesh, in gold, in
       the end-of-the-summer
Atlantic you swam in
       off a cloudless, rocky coast.
I watched you
       dip beneath the gold-flecked surface,
diving as if one could surface
       with a lifetime of precious ore.
You swam in drizzles
       of light, inside a thousand candles'

flickering, and from my shore vigil
      I wondered if such ease is born
from some imagined arrival
      or leave-taking, or merely
the genius of motion.

      Because we move so terribly
well, in many directions, I agreed
      that the pewter medallion you bought
that afternoon, *to remember it by*,
      depicted a locomotive exiting,
not entering a black tunnel.
      Driving home that evening,
at the entrance to the toll-way tunnel,
      young boys were selling roses
you bought cheap,
      and once home we found out why—
the blossoms grotesque,
      crudely wired to their pale stems.

III.
      Facts take the shape of people,
of winters, of the impassable terrain planted
      ages before the idea of hope, and yet
I do not dispense with hoping.
      I hope the mountains will bow down for you,
that this year's snows will be warm,
      that we will find each other
in another landscape,
      that when we undress in the dark
we shall we be more
      than the sum of missed callings.

Love, far-flung as any object
of desire: star, petal, celestial pail of water
        above the parched field,
let me be a fine relic to you someday,
        above the plaster ceiling, above
the strangers in #47,
        where a silver-lined cloud hangs belly-out.
Look for me
        in dried-up tumbleweed towns,
in closed-down airport bars,
        in the boarded-up winter months
of abandoned summer resorts.
        Remember how we mined through mountains
rightfully not ours to witness,
        and how approach, not arrival,
seeds the remote, benighted pearl
        of *keepsake*, of *memento*,
the tiny jewel of this unhalting journey.

Chorus

Heading toward the crux of you
is akin to sling-shooting prayers
into the wishbone
of wild geese
in the chill distance.

Even we have trouble
believing in the stoop-backed
potter without a wheel,
inventing her frantic cups
by racing tight circles
around the clay.

What does anyone expect
leaping up from inside
one woman, or two?

I place my heart in the steel drum
with the soda crackers
and rusty water
intended to tide us over
at the end of the world.

My forefinger and thumb
pose the match
whose flame
finds its faint double
in the kitchen window,
where you too are visible,
sorting spoons and peppering
the stone soup, and through
the sudden unwrung tears

brought on by the chopping
of onions, I watch you
puzzle me out, as if
we are both mouthing
the words on a dangerously empty
apothecary bottle, desperate and searching,
for syllables, ingredients, antidotes.

Evidence

A body does not experience itself falling through space.
WITTGENSTEIN

I follow the causeway's chipped trail
of clam shells smashed by seagulls
and traverse the late May marsh
and woods rung with orioles' calls.
Geese and goslings, rosa rugosa,
beach plum, rose of Sharon,
the glittering onslaught of ocean,
of misfit love guided and reguided
like the costly dredged beach
sipped slowly back out to sea.
I insist on details, plucking stones and petals,
focusing binoculars on the osprey
nesting atop the caretaker's crude platform,
the oyster boys with steel baskets
dodging the gin-drunk game warden.
The pond quietly filling to marsh
and then to grassy fields,
tells its own slow story.
I was talking to your childhood
friend about pickled pine walls, bleached
like driftwood, the inherited house
of mismatched paintings and chairs
floated in from shipwrecks.
We were hunting for metaphors
for absence, as if you were only visiting
Prague or a sick relative
and would return momentarily
with a bushel of pears.
I collect each shot glass

of thistle, of roadside, of ocean's edge and eddy's eel
bottoming out with orange ballooning hydras
near the rebuilt bridge whose pilings
don't quite reach the span,
whose architecture we never comprehended.
I'm saving my receipts—
someday they may make sense—
like those buckets left out for rainwater,
each storm drowning out the last.

Before Reining in the Moons off Mars

In the coarse-grain sands and sea glass
of this given instant, the native sea
outblues the tourist sky.

Sea dunes teeth on the horizon.
To your left ear, you raise a violet bottle
caging a dying tenor.

A man resembling young Picasso tiles a red rooftop
while balancing on his left leg.
Faith meanders a wide line, evening nears.

All the hotel windows go on holiday.
Ill-crossed lovers rowboat in the moony bay
and the idea of you, love,

Rotates, like inviolable, invisible air
gravitating inside the rolling blue-veined earth,
like the moons spinning around Mars tonight.

I inspect the contents
of a hermit crab and inhabit the shell
of a shell of an evolving archetype.

Despite the green earth's longing
to spin, despite insisting that other
grains invest it, I love this place—

And not a little, as the mother of pearl eyelids
of an oncomer's horizon approach:
sweet pause, utter retreat.

In the kitchen tonight, and out past darkness
and the superfluous lawns, who while I slept,
vanished the then-tenant peonies?

Smoke Signals

After a winter's casino-hopping
you return home for petit-fours
and find the back door padlocked
and front door open to low tide
and not the inherited loom of alfalfa.
Nothing's missing inside, but accumulated:
wicker ashtrays, a rusty typewriter,
antiquated guides to Flemish lovemaking.
Umberto, you call out, *Marco, Miss Sadie*,
whatever constellation of loveliness
misplaced. But no one's dunking French toast
on your behalf, so you grill up
the second-story bedding of absent field mice
and mix up a vodka spider.
Such surprise in every mouthful:
scented twigs, king-size bobby pins,
parchment threads of a forged love letter.
In lieu of the usual chaser of musty cognac and cigar,
you set a family of parakeets
on fire and unleash them in the house
whose bonfire you regard from a distance,
from the baled-out bay of increasing
intrigue, fashioning a seaweed necktie.
Hard to describe the light at that moment,
but it is golden and leaving,
as you rifle through the fortune cookie
hermit crabs that come when you whistle,
intent on clues regarding departure and return,
and what formula of eternities squared
you'll spend rebuilding the root cellar,
restocking the aviary, preparing
for who knows whose breathtaking arrival.

Grey Mist at Long Nook

Day-Glo fishermen wrestle the surf-
cast lines, three rods per wind-swept

wader racing between posts,
like the young internist

piloting from one well-lit
examining room to the next,

where the reader of romance novels
and the layer of bricks

wear their loose dignity
under gauze gowns whose eyelet windows

open for the laying of hands and metal probes.
If asked, the men will name

the species they're after,
the name of why they wait

in the cold dark with bubbling lanterns,
though what they're bent on

is the light that pierces
the flapping sea, that swims inside

the belly of any scaled creature
that dwells beneath the startled surface.

Drenched in fluorescence and brilliantly
blown against the grey and gritty shore,

aflame in garments meant
to distinguish them from the natural world,

they seem to be saying, We know
we're angling at the edge of our lives,

that you'll need to cull and pluck us
from the blustered swells to save us.

Already our ears are brimming
with waves, the unrelenting rumble

of traincars heavy with expectant cargo,
the raking, throat-caught gospel

of the invisible chain gang's choir.

Shipwrecked

Rocks make good companions.
Darwin's finches appetize the twilight.
I pull a mangled deck chair close
to the rich tablecloth of the sea.
Sometimes I tuck a corner
right under my chin so the starfish
and ocean liners of dancing girls
won't stain my heart's upholstery.
The mounds of inedibles
grow, so many bones: our candlelight
jewel heist in Rio, the midnight
train ride through glowing salt flats,
a balcony embrace whose patent
we applied for. Oh, how memory
armors itself, how it loops a mountain
range around this island.
Oh the rudder, oh the sail,
oh the piling of all that saved
my drowning—I made a pyre
of those remnants in good faith,
and still the smoke signals
hang out in their overnight volcano,
like bad weather that can't loophole away.

Have you received the installments
I hurry daily in the sand?
Do you sleep with police scanners?
Do you aim your hiccuping metal detector
at the stars? Do you dream underwater?

Will you miss me like salt, like rain,
like the flotsam trail snailing forever?

Mild Instructions for Travel

All the men swing like Tarzans, ripe, disrobed
to the nines, the women, scattered and few, no more ornamental
in their various epochs of undress. In magazines,
white linen suits loom appropriate for the big getaway
where a hint of marmalade dashes lapels. What
has brought us together?

In the convenience store the clerk gestures
out the front window, index finger directing
three customers all ears at the counter, explaining
location, where something *happened*, where probably
something *awful* happened, famous disappearance, triple murder,
something, or else he's giving traffic directions. In guiding
one way, does he point out the future?

In St. Mary's, the priest compares Advent
to his teenage trip to Chicago, to his girlfriend's,
when he forgot his pajamas and wondered how he might slip
from the bedroom past the family room to the bathroom
with honor. Even his poorest metaphor explains unpreparedness.
A bad priest is a true test of faith.
Can you kneel in the service, take communion,
accept more than the mystery of faith?
For whom does this skin die?

In the convenience store, everything
you need. You finger the sweets, heavens
of chocolate and Turkish taffy delights.
Tonight you stock up.
You've made a vow, you're committed,
but tonight there's still time. Tomorrow you begin
another wrinkle, but remember, newness is a constant
only in its replacement. Who says, *We live for passion*?

Do not preach. Never, never judge.
Cherubic ice sculptures have been melting
for twelve years in the back stairwell.
Tell no one.
Do not tempt strangers from their way.

In a small town on Sunday morning, there are two places
to dine early and both house the two only loves
of your life. Their faces are one, lost and indelible.
Eat nothing, or twigs. Open your collar and follow alleys
home in the cold. Make a list of your possessions: telescope,
six lamps, etc. Sell the list and move to a city
you could abandon, where you do not speak the language,
where you look familiar to anyone
only by resembling someone else.

Love in Translation

After enough history you can assume
that every other day's
the foreign language original
juxtaposed to the day you are living.
Je ne saurais plus ou me mettre:
Love is ambushing the spider.
Anioly stracone, stracone anioly:
The pastries are very fresh today.
The insomniac mailman knows,
hand-copying anonymous fan mail by the sea.
Oh, derided nuance,
I feed my face to the nonplused window,
set out on the avenue of hieroglyphics
and proposition that decoder moon.
Meteora crepuscolare: The guided tour
of famous hardware stores commences now.
Seasons and laundry.
So much has happened!
I'm sitting in the little cafe
that used to be our road map.
The waiters are stripping down
to meet the ice age.
Dearheart—*Mit uns geht das Unendliche vorbei:*
Half-price drinks, hurry up!
I'm making it all up without you.

Je ne saurais plus ou me mettre: I won't know where to put myself.
  —Apollinaire
Anioly stracone, stracone anioly: fallen angel, angel fallen
  —Polish
Meteora crepuscolare: Aurora borealis—Italian
Mit uns geht das Unendliche vorbei: Eternity is passing us by
  —Rilke

A Piece of la Femme Idéale

She can't go anywhere without linoleum
gawking up her dress or sly Cassiopeia
selling tickets down her flowery blouse.
What skunk left goose grass
and a dog-eared *Catalogue of the Universe*
by her padlocked door?
Not me—I was crafting a eulogy
for an industrious fruit fly.
I was knitting lint busts
of the village pharmacist.
Who taught those penguins
to bark Perdido in her dreams?
Who bronzed her first sneeze?
Not me—ask anybody.
I was alone in the Alibi Lounge
in Oxford, Iowa, swilling shots
of imported fertilizer.
And truth be known,
I only met her that one time,
a brief and awkward affair,
and mainly she stayed perched
on a garden ornament
in the bedroom, chirping praises and promises
overheard in a lovelorn beauty parlor,
her every hairdo flocked with infidelity.

Poem From the End of the World

I did not know
where to put
hours and rain.

I did not want
to stand where the windows
could see me.

No one had invented
the small chair
where I might fit.

So I lay awake
and counted the mountain peaks
on Chinese spoons.

I did not realize
I slept
until I dreamed,

nor that I
had dreamed
until I woke.

I did not fathom
until I stepped
outside that love,

despite us,
travels its own
chameleon road.

Where once you kneeled,
the pear tree blossoms.

Note in a Bottle

The road ends at a stone wall
or the breaking sea.
The road becomes
another road or rutted field
of daffodils where a girl imagines herself
as a road that winds
into a field fevered with flower.

Beside the sea the aquamarine shutters
of thirteen white cottages
slap September's chill like the landlocked
wings of flightless birds.
Each bleached shack holds a flower's address:
Wisteria, Primrose, Marigold, Bluebell,
Larkspur, Begonia, Petunia, Salvia,
Iris, Cosmos, Zinnia, Dahlia,
and the last, odd Crocus.

The tourists, who murmuring,
tracked moon soaked sand from Larkspur
to Primrose have sighed their good-byes,
the flapping, faded horizons
of identical clotheslines
hang empty by each boarded door.
And off to sea the frothy lace of white caps
fans out and sinks, a trail
of trains of drowning brides
who will never be more beautiful.

Like the bottle blown about the ship
the body must be built around the heart
that bears itself to open water.

There was so much I meant
to tell the wind:
Bon voyage, take pictures, keep warm.

Slowly the stars appear
like grains of rice tossed
to fertilize our dreams,
scattered crumbs
to limn a lost way home.

The road back leads ahead.
The world ends, and so begins.

III

The Cartographer's Vacation

We've all got maps in our heads:
directions to the butcher, where the woman
sets out cooling pies. The hard part's transferring
maps between people, say, helping guests find
a certain mountain pass. A written map shows
the proper turns where the nameless roads fork,

points out functional landmarks unlike
nuggets of hail on a tin roof, the wind
gossiping to a skirt, or a match struck
to illumine crickets. I scale down everything
professionally, incessantly, have mapped
every channel, each peninsula so often

I've lost count. The Canary islands, orange chips
of my thumbnail, patchwork Australia, a stain
my coffee cup formed—once they held my interest.
But now, where might I travel, exotic, assuming
I had a day of rest? From this drafting table
the Transylvanian Alps, the Amazon, are as familiar

as the arch of my foot. Might some square inch
harbor the unfamiliar, something less navigable
than my eye along the ridge of one palm's crease?
My latest theory concludes the world
is cubic, closely resembling this room where
I've plastered hundreds of maps over the floors,

eight-foot ceilings, walls and windows—the way
an antebellum bank in Columbus, Georgia (85°N, 33°E)
is today wallpapered in confederate bills.
Is my theory controversial? Take me on holiday

anywhere where one principal port, one vertiginous
face of rock's unmapped, away

from this desk where I'm accompanied
by my latest anonymous gift:
a spider from Pella, Iowa,
in a fist-sized, glass tight jar.
The spider eats nothing I feed it.
Methodically, it spins no web.

Calendar Maker

The numbers nest
like feather hats
boxed in a pale avalanche
of possibility

solo lovebirds
parked on a wire cage's
trapeze to nowhere

barkless dogs
chained in their safe pens
a double measure of love
wed atrociously
to fear

Their ancestors have seen it all
leaving charred cities
gardens of sunken bridges
promises of never again

I line the little fellows up
a portrait in innocence
while history impatient
in the wings
begins its roll call

Contrary to readers
of tea leaves and tarot
I predict nothing
but merely load the dates
like bullets
in narrow chambers

I've been known
to camouflage the future
with bactrian camels
tango prodigies in marble corridors
pastries of the month

All persiflage when
in Junction City, Kansas
a shy mechanic
yanks Miss April
from her nail
stuffs her and all her sisterhood
into a desk drawer
as an unknown woman
his wife-to-be
enters the tidy garage
with an open map
and a question
he'll take
a lifetime answering

Instructions for Writing

Eat a good breakfast.
Practice waltzing in the boiler room.
Expect flying tigers and nosy icebergs,
tides that run both ways and vultures
that address you by name.

Carry a first-aid kit.
Converse with extinct invertebrates.
Steal notes from the mimosa.
Expect to lose them.

Train your eye to distinguish
the limping, myopic guide
from the savanna
across which he leads you.

Hoard nothing.
Display your heartbeat,
the broken mirror,
the nonexistent trumpet's bleating breath.

Don't let facts
distract you
from the truth.

Include the impossibly blooming
white dahlia planted
in the blue lips
of a boy fished
from the Antarctic —

his frozen blood a logjam,
your pocketpen
the expert ice pick.

When you look ahead,
think of Lot's wife,
when you look back,
think of Lot.

Night of the Golden Gate Caper

Did you not, for instance,
on the eve of a fainthearted moon,
in a year of our Lord, from a hotel
in the south of some indistinct county,
call collect to your Great Aunt Sully,
saying, *I'm phoning from the Golden Gate Bridge?*
And did you not, for years before that,
habitually phone Aunt Sully
on nights of monodical moons or twilights
of thunderstruck calm, tantalizing or threatening
her with your whereabouts: in the gorilla cage
of a Chinese zoo, in a submarine
leaking somewhere in the Adriatic,
or from the circus tent, playing understudy
to the featured human cannonball?
Did not your Great Aunt Sully
always accept these predicaments,
never questioning your arrival in such places,
dotingly wiring favorites to your supposed locations:
boysenberry pies, daguerreotypes,
undiscovered Pepusch concertos?
When she pasted tears to the parcels,
were you not moved a little?
Is there no regret
in your silence, now, or that night,
when you must have heard the phone
go dead, imagined the cord
burying itself into flesh,
when you persuaded Sully to believe
that bridges span varying distances
and the waves galloped over her whole?

Roll Call: Between the Coups

Newly evicted, we wander the mumbo-jumbo
alleyways disguiscd as pedestrians.
Oh, endless cupboards crammed with jellies,
bottomless teacup of our desire,
we make lists of all we've lost
on sandpaper and slip them beneath the door
at The Office of Injustices when no one's looking.

Rumor goes that false and true and deep indigo
are battling it out in our yellow bathtub.
The wild boars choose all the television shows
and the local celebrity plumbers
strap candelabras to their twisted backs
and bowl all night in the corridors
with onions and deserted wine bottles.

They're all here: Houdini holed up in the fridge
and Hadrian phoning out for egg rolls,
Goldilocks swilling scotch from our best shoes
and Sisyphus mucking up the woodwork.

They've got their work cut out for them.
So who could blame them,
forgetting to invite the volunteer fire department?
How could they possibly notice the tidy bedroom
fire, the flames shooting higher and higher,
like the anxious arms of school children responding
with the one obvious answer of each day?

Who'd have thought the kindling
of our slim memories could ignite a palace?
What do they see, racing outdoors

with their smoky eyes and scarred lungs?
When they fall onto the ashed lawn and gaze up,
do they imagine the smoke trail a fishing line
lazily scanning the heavens?
Or are they like us, squinting fiercely,
sure some message is spelled out up there,
but certain too it's like following
a long-winded ace bomber turned skywriter,
whose thesis sentence eternally blurs and vanishes
long before reaching its heartfelt conclusion?

Ode to the Noodle

Little chameleon
swooning
for a tin pot
of old water,
you remind me
of our worst
lieutenant generals,
balancing
on tiptoes,
yanked tall
and then swollen
and collapsed
with greed,
curling their thick tongues
like hoodwinked nooses
down into the blue bowl—
the color of tricky sky,
well-traveled ocean glass,
the last emptied bottle
tossed overboard—
fishing for the stray
finger, for one
lovely, distracted neck.

Oh, you noodles
in noodles' clothing,
loose-limbed nourishment
of giddy schoolchildren
and wrong-faced shadows
late in December,
behind your backs
mechanical bears

still exchange exiled gifts
in holiday windows.
Across town,
the aging carpenter
unwraps gold, silver and crimson
orbs, which, hooked
to the fir's weary limbs
hang, year upon year
in early moonlight
like the careful eyes
of all those
slashed and lost.

Steam climbs
in the yellow kitchen.
Flowered paper
frays from the wall.
Little garden
of entrails, stream
of goosey icicles, godly
discards, blonde swarm
of iffy foxtails,
swank straw dinner
caught in a downpour,
look who is who.

I will season you
with fish heads,
and in the darkness
gobble you whole.

Notes from Exile

The luckiest villagers secured itineraries
for distant cliffs and promising mud flats,
paddling off in form-fitting canoes
dug from the final palmettos.
Miss M and I stayed
in the shadeless
world left behind,
adopting survivalist schemes —
fashioning hats from air-lifted tourist brochures
in Mandarin and Swahili,
making love beneath flimsy tables,
strolling arm in arm
past the grand avenue's deserted windows.
One pinkish evening we found a mannequin
and hollowed her out, made oars
of her arms and exhausted
fell asleep inside her.
In our frugal way,
we dreamed the same dream:
with bows and arrows we felled
a cumulonimbus army,
then slit and gutted
each cloud beside the water,
their bellies shimmering
like herring in a green sea.
We climbed inside the clouds
and dreamed ourselves
awake into a paradise
of here and now,
where not a soul
mistook us for sisters.

In the Air

At the beginning
of the new war
the waiter brings green curry.

Numerical figures
fall neatly in columns.

Above city hall
the mayor sights
suitcases of daffodils.

And in the desert
the boys
eat sand,
dream sand,
study the same sky
beneath which
how long ago
one grim
little man
wandered off
from where the others
warmed at the fire,
hunkered down,
and took aim.

Greek Tragedy

Even from inside the oak box
in which we've nailed him,
our schoolteacher insists
we're good children
and carries on good-naturedly
with the curriculum.
When his voice weakens,
we simply pull our chairs closer.

> *This is a prime example of situational changes*
> *in which the character remains intact.*

Finally, as dusk fills
the schoolyard, his last sighs
circle the room,
like a parakeet
in its dying cage.
And we wait
for homework, for dismissal,
our bellies growling,
the black flies
guzzling the apple orchard's
late autumn glow.

At the Front

After the generals
came and left,
suitcases bursting
with candlesticks
and virginal ear lobes,
she whispered,
Someone else is walking
inside my shoes.

There was no one
left to argue.
She heard only
the knife's blade
slicing apples,
the hand
like her hand,
the cold, sweet
crescents like pieces
of a fruit
although the armies
of black flies
had long since drained
the orchard's four seasons.

Patiently she waited
beneath the stairs
for the days
to accumulate,
for hunger
to climb down
from the trees.

The Disappearing Act

From his left ear
the magician
conjures
a silk scarf
a rainbow
of unbelievable
length
flaming
through his fingertips
to the tune
of commissioned symphonies
enlisting
the entire audience
for this trick
looping the never-
ending veil
as many times
as there are spectators
until the first
and last
stunned cries
choke methodically
in the long flame
crackling
feeding on
the dead breeze
moving
toward some village

Manual

Listen, don't blame
the clerk
of complaints and returns
when you go to him
with your bit of paper
when you shove
your bloody punctured hands
with their scratch-free warranty
onto the worn counter

Understand
he must tell you
the holes are only
imperfections
a natural part
of the fabric
that lets you know
each hand is handmade

Don't berate him
all day instead
invite him out
on his lunch break
he knows a good
out of the way place
with pickles and tweezers
on each table
he can pluck out
the lead for you
murmuring
these damn imperfections

Don't tell him
your story
don't make him choose
which side you're on
just buy him a whisky
and for god's sake
don't use words
like bullets
don't forget
his dear wife
in their tilted room
expecting a child

Safety in Numbers

All night they keep climbing
the stairs to my dim room—
the stone women, wooden women,
the women with glass noses and nowhere
else to go, whispering,
If we fell off the face of the planet,
you wouldn't hear a thing.

I let them enter
my dreams of pleasure boats
and cream sauces,
and pretty soon they're wanting
biscuits and corneal transplants,
a chauffeur from desire to desire.

They're offending all kinds
of royalty until I wake
to the dark room, silent
save for the muffled scrapings
from the graveled rooftop—
two feet near the edge,
explaining.

Community Alert

It happened in your city in a room with yellow curtains where a girl was dreaming. Reports conflict regarding the topic of her dreams that night: a school of phosphorescing fish in a moon-washed lake; a compass in a dark wood that says *you are here*.

It happened in your neighborhood while you were on holiday at the sea. Reports conflict regarding the details of his actions, but he did as he pleased, as if she were a fish or a bowl of fruit.

It happened across the street from your home, in the one-story stucco dwelling with geraniums in the window box. She had begun that day as a girl from Germany studying particle physics at the engineering school.

It happened in the room across from your room. The room behind the screen door that you watched open and close all that rainy, heartbroken spring. You watched her come and go, with bouquets of laughing friends, with groceries, with homework, with happenstance.

It happened. It was August and the window was open. If she had owned a large, brown dog named Ralph—but there was no dog. He put a plastic bag over her head. She pretended she was a fish or a bowl of fruit. She imagined she was a fish being transferred from one bowl into another.

It happened at the scene of the crime. The wooden table where you sit now with a cup of old coffee, books of Bulgarian poetry, an empty glass vase, and bills for heat and light could also, one day, be the scene of a crime. One day, all this will be evidence: the window, the door ajar or bolted tight, the places we placed ourselves, the words we chose, the silences that troop from room to room.

Tuesday Morning Coffee with Lartigue

How purely the biplanes and Didi,
snippet Zissou, elephant kites,
and chuckling Bouboutte
reside mid-air,
in the endless sepia
twilight of summer.
Here, the moment
soars, a swinging
bridge unfettered
by here and there.
A playing ball hangs
above Zizi the flying cat
and Dudu the raucous nanny,
their eager paws and palms
working the gray air
like jugglers maneuvering
the crowds of past and future.
The artist himself appears
in 1903, hastily seated
on the Pont de l'Arche,
a seven-year-old ghost
dashed into the frame
before three brothers,
so the spokes
of their timeless bicycles
peddle through him
along the gravel path
into the twentieth century,
past the park of ridiculous hats,
past rounder Zissou shelling beans
on an ocean liner,
past racehorses and a Vichy luncheon,

to the dusty Italian border
of 1927, *between the wars*,
(an idea still uninvented),
where Luigi and his somber guitar
serenade an empty touring car
under shadowless palms,
as a figure retreats
and the loose road
stammers on, undocumented,
all innocence
before the dark's unpetalling.

Echo Location

In the land of little white sticks,
how I love you. There the sky's not
a broken compass, and our bodies follow
our souls like clockwork, faithful
as two threads of an ancient, unearthed shawl.

In the land of little white sticks,
we leave our clothes on the slack line
and relegate circumstance to its proper domain:
a temporary backdrop, no more. Weather
vanes and weather mean nothing.

In the land of little
white sticks, we belong.
We step out to the cobblestone street
and under the moon's brief spotlight
hold the idea of each other
from spinning toothless off to heaven.

In the land
of little white sticks, we belong,
and when we leave,
nothing follows.

In the Cemetery

Sunday afternoon nothings so I drive
undestined until a cemetery
pulls up on the right and I park
in a closed Gulf station
that makes extra cash displaying headstones.
Across the street the cemetery's unkempt,
empty, except for brittle grass, turning maples,
and stones spaced evenly, row after row.
For no good reason,
I pace off the length: 260 strides—
an estimate because I'm distracted, reading names.
Sweeney, O'Malley, Sikes:
Irishmen who shipped their names intact,
unlike my own ancestors
who threw off -edsky, -ovitch, -insky,
like portions of themselves
as they dropped anchor.

My name's not here,
but scattered across this country:
in Pittsburgh, where a milkman
crossing an icy bridge was mugged
and drowned in three feet of water;
in Cleveland, 1943, where Nana
found her widowed sister in bed,
naked except for her head
dressed in a shopping bag;
in Brooklyn, White Fish Bay,
Bowling Green, Flint, Savannah;
en route to Sacramento
where my great-uncle Joe
slipped under a boxcar of ore;

and anonymous on another continent,
ashes lost in a countryside
I've visited only briefly,
eating sachertorte, sampling schnapps.

Today it is late in October
and a golden-leafed yew
reaches out to catch something autumnal.
I walk over people who
are not, grasses which grown
taller, or aided by an early snow,
might hide me if I stopped
to rest among them.
But what's left of my family
has enough to mourn,
and in a kitchen nearby,
someone is waiting supper.

photo by F. G. Bewer

About the Author

Andrea Cohen grew up in Atlanta, Georgia
and received an MFA in poetry from the
University of Iowa, where she was a Teaching-
Writing Fellow. Her poetry, stories, and essays
have appeared in many journals, including
Crazyhorse, *The Iowa Review*, *Orion*, and
Ploughshares. She is also a recipient of a PEN
Discovery Award for poetry. She lives in
Cambridge, Massachusetts.